Fostering the Bioeconomic Revolution ...

... in Biobased Products and Bioenergy

an Environmental Approach

Contents

Executive Summary

The Revolution. A revolution spurred by expanded use of renewable biobased products and bioenergy will have enormous impacts in the 21st century. Today, America depends on biomass to provide 3 percent of its energy and more than 300 billion pounds of products annually. Yet, propelled by advances in biological and physical sciences and engineering, America has the opportunity to greatly accelerate its use of biomass.

The Importance. This revolution will have important impacts on our:

- *Economy* — stimulating growth, especially in rural, farm and forest economies, and our industries.
- *Environment* — by reducing emissions of carbon dioxide and airborne pollutants, reducing soil erosion, sequestering carbon, protecting water supplies and quality, and increasing the diversity of crops and products.
- *Energy security* — by increasing domestic production and reducing our enormous dependence on foreign sources of fossil fuels.
- *Competitive position* — opening up new technologies, industries, and export opportunities.

The Consensus:

- The National Academy of Sciences recognized the predominant impact that biological sciences will have on industry and technology in the 21st century.
- The President proclaimed the importance of biobased products and bioenergy by establishing a national goal to triple their use in the United States by 2010.
- The U.S. Congress acknowledged the exceptional potential of biobased products and bioenergy by enacting the Biomass Research and Development Act of 2000.

- Industry, government laboratories, academia, and others recognized the importance of biobased products and bioenergy and are developing strategic planning documents including visions and technology roadmaps.

The Challenges:

- *Science and technology* — we need to advance science and develop technologies to overcome difficulties posed by the complexity of biomass resources and processes.
- *Policies* – we need to coordinate government policies to meet the national goal.
- *Markets* — we need to accelerate the commercialization of new and emerging technologies and products to meet the national goal.
- *Environment* — we seek to ensure that new technologies and increased use of biomass will not adversely affect our land, water, air, and public health, but rather provide environmental benefits.
- *Education and Outreach* — we need to provide American institutions, industry, farmers, landowners, and the public with information that will help them understand biobased products and bioenergy and make wise decisions in their application. An informed public is critical to the decision-making process.
- *Integration* — we need to coordinate efforts across federal and state programs, educational organizations, businesses, and technologies to encourage the growth of an integrated industry.

The Plan. This plan describes the federal strategy for meeting these challenges and attaining the national goal of tripling the use of biobased products and bioenergy by 2010. It is built on a strong foundation of science and technology, policy, and market goals, and strategies. The goals are summarized below and detailed in the pages that follow.

Technology Goals

- Reduce by two- to ten-fold costs of technologies for integrated supply, conversion, manufacturing, and application systems for biobased products and bioenergy by 2010.
- Accelerate commercial readiness and acceptance of integrated biobased products and bioenergy systems for fuels, heat, power, chemicals, and materials.
- Assess environmental and ecosystem impacts of, and enhance the benefits of, biobased products and bioenergy systems at all stages of development.
- Foster innovation-driven science of biomass feedstocks, biobased products, and bioenergy and quickly incorporate these scientific results in the relevant technology-development activities.

Market and Public Policy Goals

- Coordinate policies to achieve early market adoption of biobased products and bioenergy and create demand for biobased products and bioenergy.
- Increase federal government purchases or production of bioenergy to 5 percent and relevant biobased products purchases to 10 percent by 2010.
- Facilitate tripling the use of emerging biobased products and bioenergy by 2010 in a manner that is consistent with federal resource conservation and environmental policies.

A Time and Consensus for Action

We are entering the 21st century at the vanguard of a revolution in biology. The National Academy of Sciences has stated that the "Biological sciences are likely to make the same impact on the formation of new industries in the [new] century as the physical and chemical sciences have had on industrial development in the [last] century." From biological resources we can derive products as diverse as food, feed, building materials, paper, clothing, fuels and lubricants, chemicals, heat and electricity, and much more. Because of scientific breakthroughs that are occurring in the biological sciences and in process engineering, we can look forward to a great expansion of biobased products and bioenergy.

Products and energy based on biological processes and resources will help provide the nation with a healthy, growing, productive, and sustainable economy. They will help relieve environmental pressures, improve our rural economy, reduce our balance of payments, and enhance our energy security. Furthermore, they will help the United States maintain its global economic leadership.

Recognizing this potential, the President issued an Executive Order on Developing and Promoting Biobased Products and Bioenergy (64FR 44639 – Executive Order 13134) on August 12, 1999. The Order and Executive Menorandum call for coordination of federal efforts to accelerate the development of 21st century biobased industries and to triple the U. S. use of new and emerging biobased products and bioenergy by 2010. This will involve development of a comprehensive national strategy, including the following:

• Research and development (R&D) and demonstration.

• Education and outreach.

• Supportive policy actions.

• Incentives to stimulate the creation and early adoption of technologies needed to make biobased products and bioenergy competitive with fossil-fuel-based alternatives.

Following the President's lead, the U.S. Congress enacted The Biomass Research and Development Act of 2000 (Title III of the Agricultural Risk Protection Act of 2000, H.R. 2559 – Public Law 106-224) on June 20, 2000. This Act complements many aspects of the Executive Order. It directs the Secretary of Agriculture and the Secretary of Energy to jointly submit a report that describes the research and development goals of the initiative, and the resources required to reach those goals.

As required by both the Executive Order and the Act, integration of federal efforts is underway. Historically, the Department of Agriculture (USDA) and the Department of Energy (DOE) have performed more than 90 percent of the federal R&D on biobased products and bioenergy. The National Science Foundation (NSF), the Environmental Protection Agency (EPA), and the Department of Commerce (DOC) also have programs and projects

The National Vision
Provide technology and market development for a strong industry that will create new economic opportunities for rural America, protect and enhance our environment, strengthen U.S. energy and economic security, and bring improved products to consumers.

The National Goal
Triple the U.S. use of emerging biobased products and bioenergy by 2010. (See Figure 1)

The Federal Mission
Accelerate the research, development, delivery, and adoption of technologies needed to make biobased products and bioenergy competitive in national and international markets. Concurrently, promote the efficient use of our renewable resources and protect the environment.

The National Partnership
Many different industries, academia across multiple disciplines, government across several agencies at federal, state, and local levels, and many non-government organizations will cooperate to make the national vision a reality.

in support of biobased products and bioenergy RD&D. These, and other federal agencies, are part of the Biomass Research and the Development Board. The mission of the Board is to coordinate federal efforts in planning, funding, and R&D, and to accelerate the integration of a biobased U.S. industry and the development and commercialization of biobased products and bioenergy.

The Board's endeavors have already resulted in several coordinated efforts, including:

• Report to the President on Executive Order 13134.*

• Development of a national vision statement by industry and other stakeholders facilitated by DOE.

• Competitive Fiscal Year (FY) 2000 solicitations for R&D on biobased products and bioenergy will provide for 100 awards ($50 million) covering a wide range of topics, including integration (DOE and USDA).

• Establishment of the Biomass Research and Development Act Technical Advisory Committee.

• A joint budget of $289 million for FY 2001 R&D was requested for DOE and USDA, of which $230 million were appropriated. The overall DOE and USDA R&D appropriation is at the same total level of the federal FY 1998 inventory shown in Figure 2 (FY 1998 is the baseline year of the inventory, including the disposition of grants by technical area by agency).

• In FY 2001 and FY 2002 the USDA will provide up to $150 million annually to encourage increased purchases of commodities by companies for expanded production of biofuels.

• Expertise is being leveraged across agencies and between government and industry.

The strategy presented in this document is based on programs, projects, and plans from the relevant participating agencies, on input from a wide range of public and private stakeholders, and on planning documents produced for a variety of biobased technologies. It presents a high-level summary of the emerging national strategy. It provides the first integrated approach to policies and procedures that will promote R&D and demonstration leading to accelerated production of biobased products and bioenergy. This summary is backed up by more detailed companion documents including:

• The federal inventory of activities concerning biobased products and bioenergy.

• Detailed goals, metrics, milestones, and resources.

• The relevant program plans of the government agencies involved (See appendix I).

---

* The references used to prepare the Strategic Plan are cited in the Report to the President (2000) at http://www.bioproducts-bioenergy.gov/.

Biobased Products and Bioenergy Goals

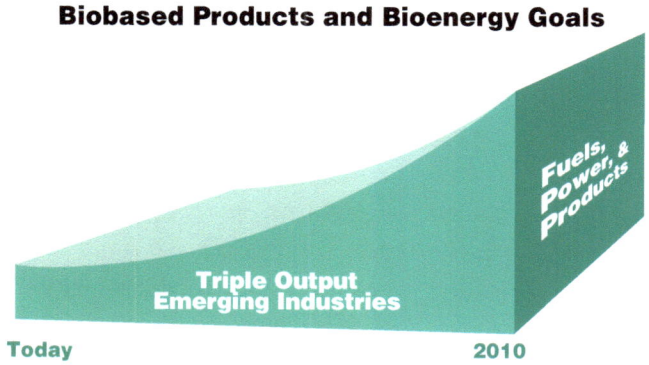

Current baseline for emerging industries:
• ethanol, 1.5 billion gallons
• biodiesel, 6 million gallons
• electricity, 60 billion kWh (from 10 thousand megawatts of capacity)
• emerging products, 10-15 billion pounds (5 - 7.5 million tons).

Figure 1. Biobased Products and Bioenergy Goals: Consistent with federal resource conservation and environmental policies, triple output of emerging industries in fuels, power, and products and facilitate an increase in efficiency of use in mature industries, with special attention to application of new and emerging technologies. (Note: triple refers to specific fuels, power, and products output or any combination).

Fiscal Year 1998 Federal R&D Funding – $253 Million

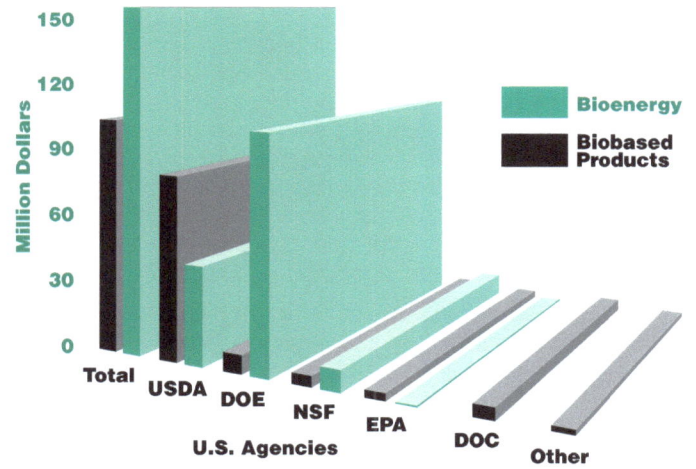

AREA	Funding in millions of $	Percent of total
Bioenergy Science	$53.5	21.1%
Biofuels	$47.5	18.8%
Crosscutting Science of Biobased Products	$34.8	13.7%
Green Chemicals and Plastics	$34.7	13.7%
Biomass Resources	$24	9.5%

AREA	Funding in millions of $	Percent of total
Biopower	$23.7	9.4%
Natural Structural Materials	$17.6	6.9%
Natural Fibers	$12.3	4.9%
Integrated Assessments	$5.2	2.1%

Figure 2. United States Federal investment in biobased products and bioenergy is based on Fiscal Year 1998, the year for which the detailed description of projects could be compiled along with the respective funding levels.

Challenges and Opportunities

Carbon-based products from both fossil and renewable sources play a critical role in our economy. They are ubiquitous in our fuels, in our packaging and clothes, in our vehicles and homes, and in our work and leisure activities. Renewable resources such as trees, grasses, and crops were the primary source for such products until the mid-19th century. Then gradually, as our understanding of chemical, geological, and physical sciences and engineering progressed, and the demands of industrialization grew, fossil fuels began to replace renewable resources and became the dominant raw material for energy, chemicals, and products in the 20th century. Today, the cost and security risks of fossil fuel imports, environmental concerns with increased pollution and with global greenhouse gas emissions from fossil fuel use, the desire to improve the rural economy, and breakthroughs in biological sciences, technologies, and processes have renewed our interest in biobased products and bioenergy.

These advances in the science and technology of biobased products and bioenergy present a great opportunity to expand our renewable carbon-based industries while simultaneously moving toward a more environmentally sustainable economy. Biobased products and bioenergy offer considerable environmental benefits. They often require less energy to produce than the fossil and inorganic products they replace. Biobased products often reduce waste as well as improve air and water quality relative to the products they replace. In addition, biobased products can sequester large amounts of carbon while adding little if any net carbon emissions to the atmosphere.

However, biological feedstock growth, processing, and use can pose environmental challenges as well. Changes in land use, pesticide, fertilizer, water, and other requirements could carry risks to habitat, public health, diversity, and air and water quality. To ensure that this initiative maximizes environmental benefits and minimizes potential risks, the environmental benefits and costs will be assessed at all stages, from planning to commercialization.

Biologically-based technologies can often be more versatile than fossil-based ones. Because of the great diversity of biomass resources, we can also create valuable, novel products with exciting new properties. Whatever products we can make with fossil fuels, we can make virtually identical or better ones from biomass.

However, the challenges of developing new biobased products and bioenergy are formidable. Biomass resources are more complex and harder to process than petroleum or coal. A large number of production, conversion, and utilization technologies are possible. There are many ways to integrate individual technologies. Because of the fast pace of progress in biological sciences and technologies, the number of alternatives will increase rapidly. Some companies could oppose efforts to challenge the status quo. Others, however, are embracing this life sciences-driven business opportunity.

We can overcome many of these challenges using integrated, systematic R&D—to drive innovation, reduce costs, and to help U.S. industry gain a leading edge in the biology revolution now beginning.

Timely investment by public-private partnerships will help us gain that edge. For new and emerging biobased products and bioenergy to more quickly penetrate the marketplace we need to accelerate R&D and demonstration, foster commercialization, facilitate market development, and increase efforts in education and outreach. To create biobased products and bioenergy, we must develop integrated plant science, technologies for crop and forest production, and technologies for converting biomass into fuels, materials, chemicals, electricity and heat.

These investments will enable us to explore a wide variety of pathways through which we will be able to produce green plants, transform these plants into cost-competitive products and energy, and increase the pace at which the products and technologies penetrate the markets. Timely investment will further enable us to more readily incorporate environmental principles throughout the plant growth-conversion-utilization cycle.

Key Examples

Sugars and biosynthesis gas from biomass are two examples of building blocks from which we derive many other products. We support these building blocks by developing critical technologies such as biomass hydrolysis and gasification. With these critical technologies we can unlock a wide range of products — just like thermal and catalytic cracking did for petroleum in the 20th century, when we learned to refine crude petroleum into the fuels and more than 70,000 chemicals used today.

Sugars. Sugars could be as important to the bioproducts industry as ethylene is to today's petrochemical industry. The chemical industry uses ethylene as a starting material to make thousands of consumer products, including poly-ethylene bags and polyethylene terephthalate (plastic bottles for soft drinks). Analogously, with sugars derived from lignocellulosics found in trees, grasses, or residues from agriculture crops, we can use fermentation and chemistry to make hundreds of products including:

- *Alcohols,* such as ethanol, glycols, and sorbitol. Ethanol is used as an oxygenated fuel that helps reduce toxic air pollutants and increase gasoline octane numbers. Glycols are used for making antifreeze, brake fluids, and solvents. And sorbitol is used in adhesives, as a soften-ing agent, and as a sweetener.

- *Acids,* such as lactic acid, which is used for preparing cheese, soft drinks, and other food products. It is also developing into a starting material for biodegradable plastics.

- *Polymers,* such as xanthan gum, which is used as a food-thickening agent and as a gel in toothpaste, medicines, and paints.

Figure 3 illustrates how this initiative could reduce the cost of sugars faster than the current plan to create cost-effective starting materials for biobased products and biofuels.

Biosynthesis Gas. Biomass gasification uses heat to convert solid biomass into a biosynthesis gas—which is primarily a mixture of carbon monoxide and hydrogen, with carbon dioxide, water vapor, and small amounts of tar. Once this biosynthesis gas is cleaned of tars we can use it to make:

- *Electricity.* We can use biosynthesis gas in advanced turbines or in fuel cells to produce electricity at more than twice the efficiency of today's combustion systems. By using gasifica-tion technology to replace aging power and heat-generating equipment, the pulp and paper industry could become energy self-sufficient and could even export up to 30,000 megawatts of electricity by 2030.

- *Alcohols.* One such alcohol is methanol, which is used in antifreeze solutions, to produce formaldehyde (used in plastics, germicides, and fungicides) and other chemicals, and as a high-performance fuel.

- *Acids.* These include acetic acid, which is used in photographic films, textiles, vinyl plastics, and polyesters.

- *Clean hydrocarbon fuels.* Using a Fischer-Tropsch process, synthesis gas is reacted on a

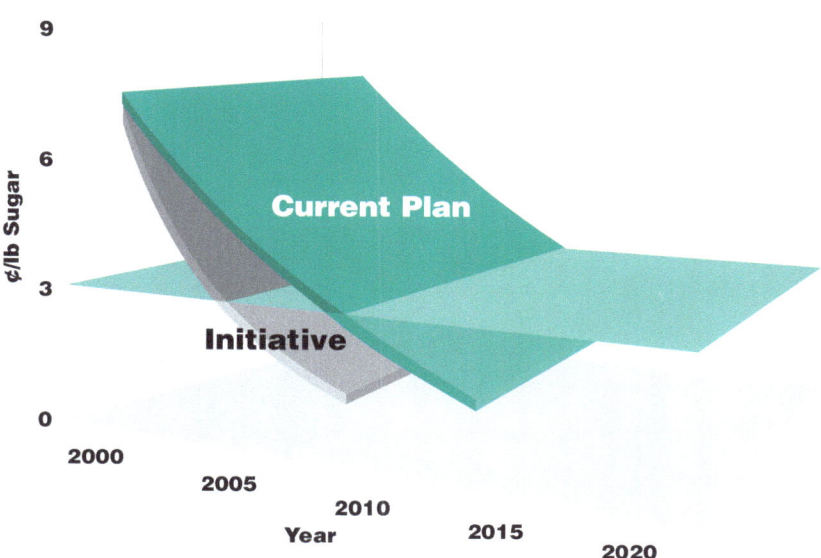

Cost Reduction Targets for Sugar

Figure 3. Although R&D helped reduce costs of sugars from lignocellu-losic biomass by a factor of six since 1980, further reduction in cost is needed for widespread use of sugars for fuels, chemicals, materials, and other products.

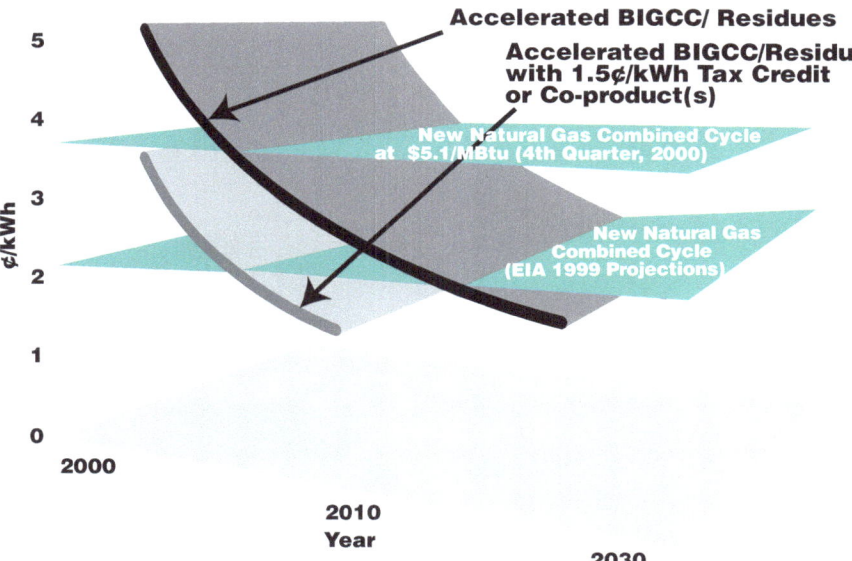

Projected Cost of Electricity from Biomass Residues Integrated Gasification Combined Cycle (BIGCC)

Figure 4. Accelerated R&D can reduce the cost of electricity from bio-mass residues compared to today's lowest electricity cost from new nat-ural gas using combined-cycle. Biomass electricity costs could be reduced even further and become competitive earlier if coproducts are developed quickly. Alternatively, costs can be reduced earlier if pro-duction tax credits are maintained over a sufficiently long period of time to justify private-sector investment, new technology development, and adoption. (EIA is the Energy Information Administration) (Note: costs assume municipal financing).

catalyst to make gasoline and diesel fuels. When this process is used with biosynthesis gas, it produces fuels that are free of sulfur.

- *Other products.* This includes many products we currently make with fossil resources, including some plastics.

Figure 4 shows how we could use a variety of strategies to reduce the cost of biomass-generated electricity to be competitive with electricity generated with natural gas using combined-cycle turbine technology. Specifically, if we were to use low-cost biomass residues in accelerated gasification technology development and couple it with combined-cycle turbine technology, biomass electricity would become cost competitive by 2010. This projection assumes the Energy Information Administration's (EIA) 1999 natural gas cost scenario holds. In the fourth quarter of 2000, however, we have seen natural gas price nearly double that of the price levels in the EIA projection (also shown in the figure). Alternatively, if gasification technologies are developed to make a coproduct and electricity from biomass residues, cost competitiveness with low-cost natural gas could be reached earlier. If we add production incentives into the mix, biomass electricity could become competitive before 2005 and would help increase private-sector investments in these high-efficiency systems. This initiative will enhance the realization of early market penetration strategies.

Using biomass and its derivatives as starting materials and intermediates that can be refined into a variety of products are functions of a **biorefinery**, a concept that is analogous to petroleum refineries. Biorefineries use complex processing strategies to efficiently produce a diverse and flexible mix of fuels, electricity, heat, chemicals, and material products. Today, biorefineries are present in our agricultural and forest products industries. These industries have created systems that provide food, feed, fiber, materials, and some chemicals and energy as heat and electricity. A variety of products, a few illustrated below, are emerging.

Biobased Products. Two of many important classes of biobased products are plastics and surfactants.

- *Plastics.* The chemical industry today produces more than 80 billion pounds of plastic products annually. The great majority of these products are derived from fossil resources. Biomass resources account for only a small percentage, many filling high-price, value-added applications. Accelerated R&D could help biomass-derived plastics in two ways. First, it could help lower the costs of a wide range of biobased plastics, which would then enable these plastics to penetrate high-volume applications that could offset increased petroleum-based plastics production. Second, it could help continue the development of a slate of high-quality plastics with unique properties, which would penetrate many markets based on their performance. A variety of important specialty plastics could be sold at premium prices. Figure 5 shows the effect that accelerated R&D could have on the biobased plastics market—through the development of a wide array of biobased plastic products at various prices. The biobased plastics market could reach 3 billion pounds or more by 2010.

- *Surfactants.* These compounds are used primarily in detergents, cleansing products, and in medications. About 35 percent of today's surfactants are derived from biomass resources. Biobased surfactants include low-price to high-value products. Accelerated R&D would enhance the market for biobased surfactants in all price ranges.

Given what R&D could do for just these two classes of biobased products, imagine the possibilities when applied to biomass natural fibers, wood, starches, cellulose, and lignin polymers. And when we include R&D to make plants that

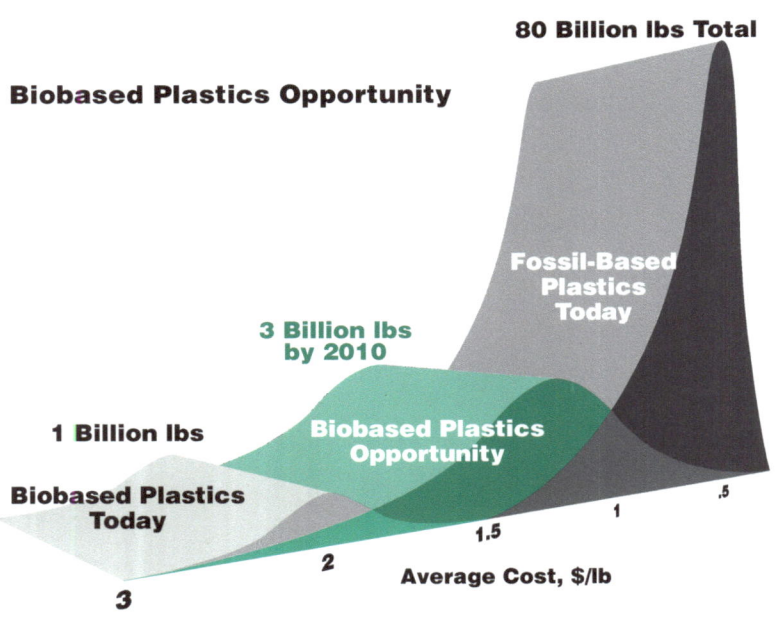

Biobased Plastics Opportunity

80 Billion lbs Total

Fossil-Based Plastics Today

3 Billion lbs by 2010

Biobased Plastics Opportunity

1 Billion lbs

Biobased Plastics Today

Average Cost, $/lb

.5   1   1.5   2   3

Figure 5. Pictorial example of today's costs and volumes (distribution) of biobased and fossil-derived plastics. In the future, these distributions could change markedly, as shown in the center distribution curve. Biobased products could generate 3 billion pounds or more of products covering a wide range of costs and volumes by 2010 and significantly offset petroleum use.

directly produce biobased products with new and improved properties, the possibilities multiply.

A major advantage of expanding biorefineries is that the starting materials (biomass) are grown in the United States by fixing carbon dioxide from the atmosphere through photosynthesis. Biomass, however, is a complex resource. We must pay attention to the environmental and ecological consequences of larger-scale biomass production and use and design environmentally sound processes and products

Biomass Feedstock. Finally, the common starting point for all these products is the biomass feedstock—green plants. Research and development are critical for increasing the supplies of sustainably grown crops and for reducing the cost of biomass so that it will become America's green petroleum of the 21st century.

Feedstock research includes plant science, understanding and using plant productivity factors in each region of the U.S., and the growth and selection of plants with specific attributes. As selected plants move into development, cost-effective, regionally adapted, and environmentally beneficial crop production methods are needed. Improved crop harvest, handling, storage, and transportation methods could further assure reliable biomass supplies. In this way a broad range of cost-competitive products can be made from biomass feedstock. Figure 6 shows one example of increased competitive supplies of energy crops if plant science and technology development were accelerated.

Projecting into the future, a wide array of new biobased products and bioenergy could position our nation to reap the benefits of the biological revolution anticipated by the National Academy of Sciences and others, and to become a global leader in developing sustainable economic growth based on biobased products and bioenergy.

In the two sections that follow, we analyze the current situation in biobased products and bioenergy and indicate how this initiative could spur progress and help America meet tomorrow's needs.

Biobased Products

TODAY

Renewable resources provide more than 300 billion pounds of carbon-based products yearly, primarily derived from forest products. This number does not include food or feed. The chemical industry also produces about 300 billion pounds of carbon-based products, the majority of which are derived from fossil resources. Together, these 600 billion pounds of materials are used in a wide variety of consumer products for home, office, leisure, trans-

portation, communications, and industrial applications.

Biobased products can be found in a number of product categories along with products derived from fossil and inorganic resources. In many cases, the lowest cost product is the market winner; and in other cases, the product performance determines market dominance. In many cases, biobased products and fossil-derived products are used together in consumer products. Examples of the main categories of products are shown on Table 1.

The established companies of the pulp, paper, and packaging, and of the wood products industries have thousands of industrial plants producing a myriad of products. In addition, there are about 250 companies that produce a wide range of biobased product lines (product categories 3-16 of Table 1). The companies include both large and small businesses, many of which are still developing their first products. The development of many of these companies was facilitated by the federal mandated use of recycled products and by the low cost, availability, and regenerability of agriculture and forest products starting materials.

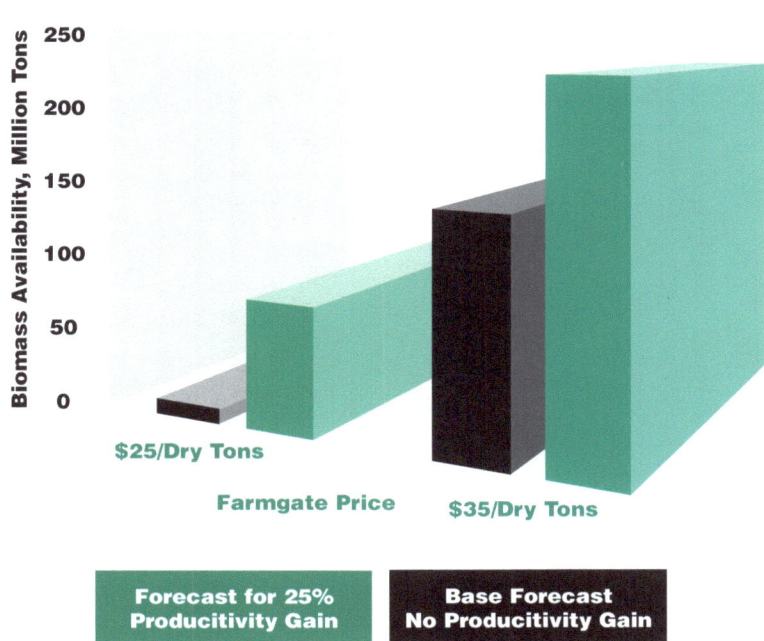

Figure 6. Plant science research is required to increase productivity and reduce uncertainty of biomass crop supplies. Productivity increases could lead to more supplies of energy crops that are economically competitive by 2008.

Figure 7 shows some examples of locations of manufacturing plants of biobased products using the system of Table 1 (pulp and paper companies are illustrated with the energy products in Figure 8).

TOMORROW

For bioased products, domestic and international opportunities abound. These include opportunities for increasing the production of current and new crops; opportunities for improving the efficiency and cost-effectiveness of technologies for converting these crops to myriad products; and opportunities for developing advanced technologies that will lead to new products with novel properties. Before long, for example, we could witness the emergence of new biobased commercial and industrial chemicals, pharmaceuticals, and products whose large capacity for sequestering carbon could benefit the environment. These developments depend on advances in biology, plant genomics, computational science, biotechnology, conversion, and transport; and on an improved understanding of natural products in their environment and our increased ability to tailor products made by green plants.

TABLE 1 Biobased Products Categories and Examples

Number	Product Category	Examples of Biobased Products
	Established Biobased Industry	
1	Paper and packaging	Writing papers, newsprint, magazines, and packaging cartons
2	Wood-based composite materials and structures	Lumber, plywood, flooring, furniture, laminates, engineered wall systems, wood/polymer and structural composites, and lignin-based polymers
	Emerging Biobased Industry	
3	Plant-based plastics and polymers and films	Polylactide plastic, starch biodegradable polymers, spider silk polymers
4	Lubricants and functional fluids products	Biodegradable soybean oil-derived lubricants, used grease-refined
5	Inks	Soybean-derived inks
6	Enzymes	Cellulase for orange juice clarification and stone-washed jeans, amylase for corn industry, enzymes for nutrition enhancement, novel property enzymes
7	Renewable alternative fiber papers and packaging	Kenaf, milkweed, and other agriculture products used for fibers, packaging, and products
	Shared by Biobased Companies and other Manufacturing Industries	
8	Absorbents, adsorbents, and masonry and road materials	Odor control, spill absorbents, animal bedding, pet litter, biocement support, roofing, insulation, road oil, and asphalt
9	Adhesives and bonding products	Sealants, glues for building products, glues for envelopes, wall paper adhesives, soy-based adhesives, marine glues
10	Biocontrol products	Soil amendments, such as topsoil, aggregate, and enrichment, fertilizer and pesticide carriers
11	Solvents, chemical intermediates, and cleaning agents	Methyltetrahydrofuran from levulinic acid, methanol from sythesis gas, cleaners, conditioners, and surfactants
12	Coatings and paints	Paints using cellulose-derived water soluble polymers
13	Cosmetics and personal-care products	Biobased products in toothpaste, lotions, and shampoos
14	Landscaping products	Decorative bark, railroad ties
15	New fibers, fillers, yarn, and insulation	Cotton fibers and rayon (cellulose derivative) textiles. New insulation using cotton processing trash and recycled textile fibers, filler for auto fenders, and panels for vehicle liners
16	Pharmaceuticals and veterinary products	Taxol for cancer treatment

There will also be opportunities for agriculture and forestry producers and processors to increase their income. Much of this will come about because of advances in the science and technologies of plant management and utilization—advances that will provide significant increases in feedstock productivity and in new systems for producing agricultural and woody crops. Additional income will also result from farmers moving up the value-added chain by establishing farmer-owned processing and manufacturing enterprises. At the same time, by increasing the utilization of biomass and residues we will improve the health of our agricultural lands, and of our forests, and we will derive other environmental benefits.

We are starting to see some of these opportunities emerge even today. Traditional chemical polymer manufacturers are forming new ventures with traditional agricultural processing companies to develop and commercialize biodegradable polymers from glucose-derived polylactide and new fibers from microbially produced chemicals. Synergies like this will continue to flourish, with companies that form these ventures being able to leverage each others' expertise and effectively explore biobased plastics and other areas together. In addition to the large processing plants resulting from new ventures, we could also witness the creation of small processing plants in rural America, which would produce new forest- and crop-based products and provide many new jobs.

Much of the initial and final processing of bioproducts will occur close to where the biomass feedstocks are produced. That means new economic opportunity for the farmers who produce the feedstock and for farmer-owned cooperatives engaged in processing biobased products.

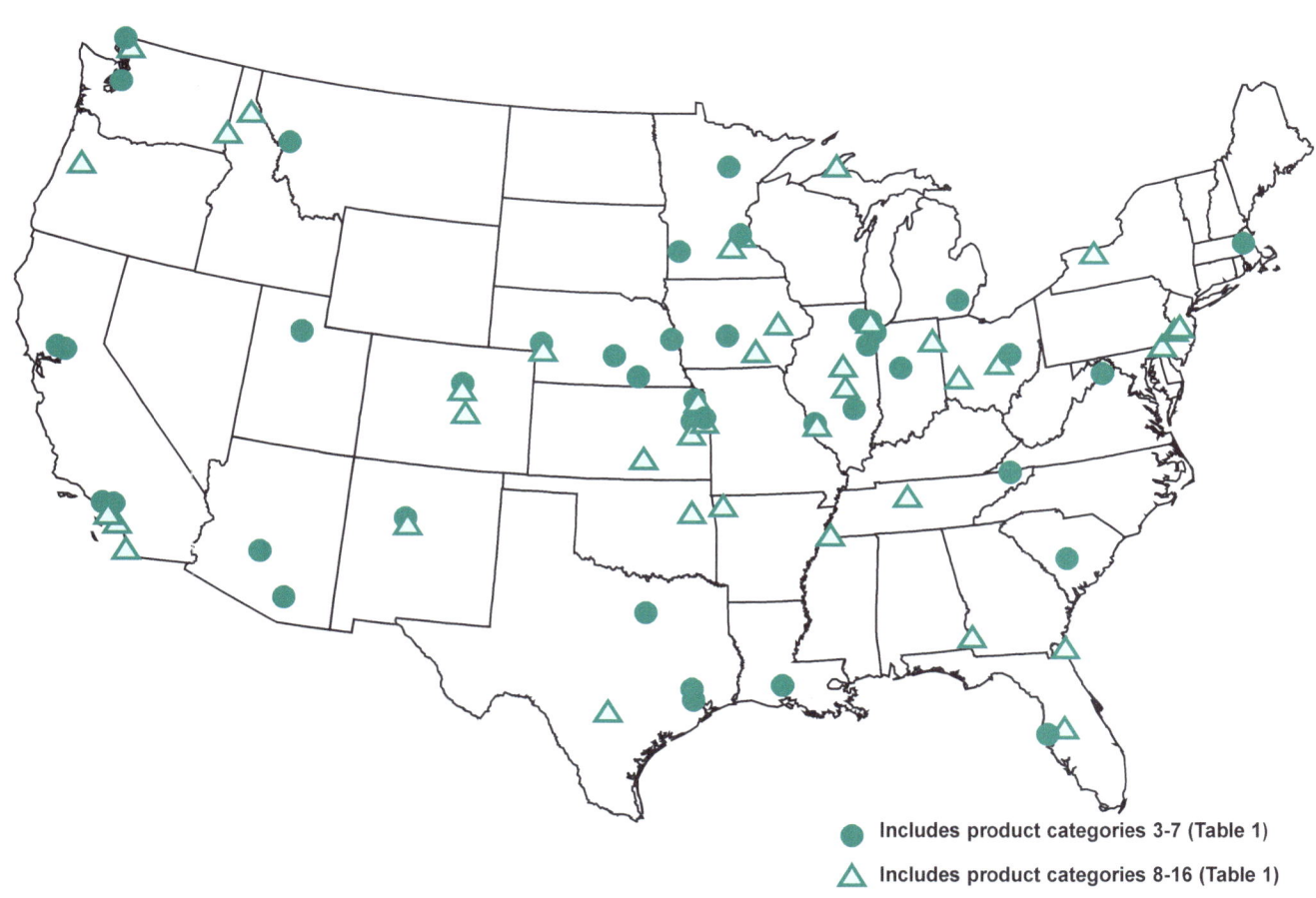

● Includes product categories 3-7 (Table 1)

△ Includes product categories 8-16 (Table 1)

Figure 7. Map of the United States with examples of locations of facilities manufacturing biobased products. About 134 manufacturing plants are included, and a single symbol often represents multiple plants.

Bioenergy

TODAY

Biomass resources supply the United States with 3 percent of its primary energy. From this primary biomass input, nearly half is lost in conversion processes. The other half is consumed in the form of heat, electricity, and liquid or gaseous fuels. Major bioenergy uses by sector include the following:

- *Buildings.* About 25 million homes use wood for primary or supplemental heating, and wood provides 10% of total residential heating in America.

- *Electricity.* Biomass residues, municipal wastes, and landfill gas are used to generate heat and 60 billion kilowatt-hours from 10 thousand megawatts of electric power (nearly 1 percent of the generating capacity in the United States). The electricity derived from the biomass residues avoided 7 million tons of carbon emissions per year because of diversion of biomass from landfills and their resulting emissions.

- *Industry.* Biomass process streams and residues provide 56 percent of the electricity and heat used by the pulp and paper industry and 75 percent of the electricity and heat used by the solid and engineered wood products industries and composites.

- *Transportation.* Ethanol (primarily derived from cornstarch) accounts for 0.4 percent of liquid fuels and is provided as an ethanol-gasoline blend in 3 percent of U.S. gasoline supplies. Of the 35 billion gallons of diesel consumed in 1998, 6 million gallons were biodiesel.

During the past two decades, federal regulatory, tax, R&D, and other policies, together with complementary policies in many states, have significantly increased bioenergy use. Opening the regulated electric utility industry to external supplies of electricity through the Public Utility Regulatory Policy Act (PURPA) enabled the development of independent suppliers of biomass-generated power. Federal and state tax incentives helped increase U.S. biopower generation more than three-fold during the past two decades, with most of this increase occurring between 1984 and 1988. Policies that supported these increases included performance-based incentives such as guaranteed prices for energy or energy production credits. Utility deregulation continues today and offers opportunities and challenges to the biopower industry

Since 1980, ethanol production from cornstarch increased by a factor of ten. This was brought about partly by our need to reduce air pollutants in many of our cities in response to the Clean Air Act, and partly by reducing motor fuel excise taxes on fuels derived from ethanol.

As a result, in the United States during the past 20 years investors installed 730 bioenergy electric-generating facilities, 58 ethanol production plants, and many more plants that generate heat only (Figure 8). To put these numbers in perspective, the U.S. biopower levels are higher than electricity produced and used from all energy sources in individual countries, such as Portugal, Greece, the Philippines, or Chile.

TOMORROW

There are significant opportunities to improve the efficiency of converting primary biomass to convenient electricity, heat, and transportation fuels. Today's bioenergy system is a composite of technologies where individual efficiencies range from 15 percent to 60 percent. When computed as the ratio of output to input energy, the average efficiency across all of these uses is between 40 percent and 50 percent. By incorporating advanced technologies, such as for combining heat and electricity production, and for producing liquid fuels, industry could raise the average efficiency of the composite U.S. bioenergy system to between 60 percent and 80 percent. As an example, by combining heat and power applications or by using advanced technology, we could double the efficiency of today's stand-alone direct combustion plants.

For bioenergy to become more competitive and more widely used in the United States, increased use of combined heat and power would help. But even greater increases in efficiency are needed. And this can be achieved only through advanced technologies like advanced biomass gasification coupled with advanced turbines in combined cycles, possibly with fuel cells (which could reach conversion efficiencies of 60 percent or higher). We must reduce the technical and commercial risk of these technologies so that industry will adopt them in an expeditious manner as they replace capital equipment and develop additional supplies.

Some bioenergy deployment is compatible with the current fossil energy infrastructure. Gasification of biomass is a feasible way to combine biomass energy with coal at very high proportions of biomass. It is also a way to couple biosynthesis gas (processed to pipeline quality) to increase natural gas supplies for distribution and home applications. Cofiring solid biomass residues along with coal directly could lead to biomass efficiency increases with relatively low capital investments.

During the 1980s and 1990s, the residue-based industry invested $15 billion to develop 7,000 megawatts of electric power production capacity. This created 66,000 jobs and a net annual income of $2 billion. By implementing advanced technologies and by coupling with the fossil fuels infrastruc-

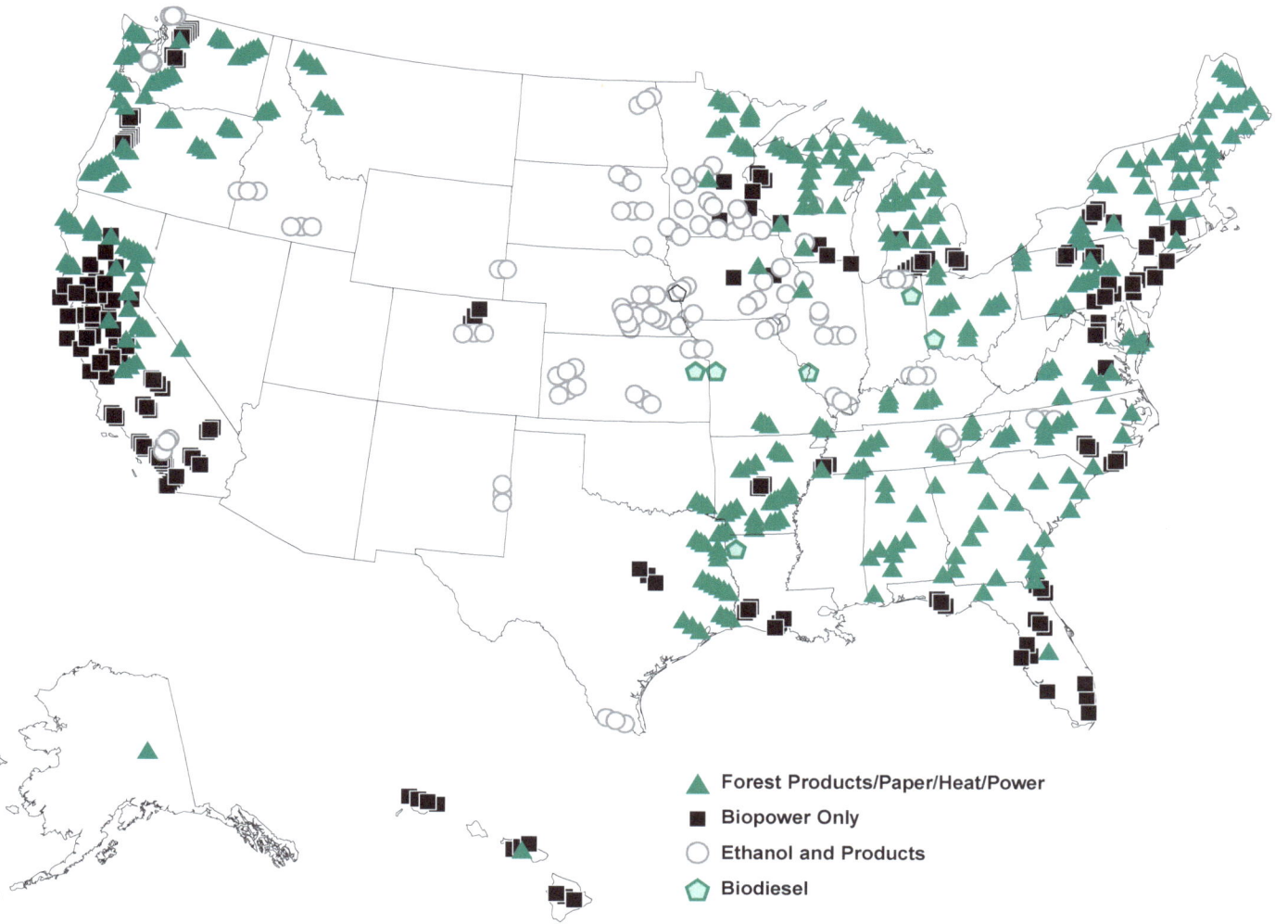

Figure 8. Map of the United States with 440 examples of locations of facilities (county level) in forest products industries that integrate heat, power, and products; biopower plants that generate electricity only; and plants that primarily produce corn-derived ethanol; and plants that produce biodiesel. Note, that a single symbol often represents multiple plants.

Legend:
▲ Forest Products/Paper/Heat/Power
■ Biopower Only
○ Ethanol and Products
⬠ Biodiesel

ture, the biopower industry could grow by another factor of 3 in the next 10 years. If this were to be the case, then the industry could help create nearly another 200,000 jobs. In fact, according to its own technology roadmap, the biopower industry has the goal of tripling its year 2000 output, while doubling its use of primary biomass, by 2010. The industry envisions penetration of their technologies in domestic and international markets, initially coupled with increased use of biomass residues and wastes.

In the liquid-fuel infrastructure, there is an opportunity to increase production of ethanol and help maintain oxygenated fuels in reformulated gasoline. These oxygenates have been shown to help improve the air quality in the regions with high air pollution to bring them into compliance with the requirements of the Clean Air Act and its amendments. There are, however, some evaporative emissions and air pollutants from renewable oxygenates that need to be mitigated.

Methyl tertiary butyl ether (MTBE), the most widely used gasoline additive to improve air quality, is being phased out in California and key northeastern states because of water contamination from leaking gasoline storage tanks. This could help increase demand for ethanol by as much as 1.2 billion gallons per year over the current demand of 1.5 billion gallons per year (a factor of 1.8). With this added level of production, which could go even higher, the balance of trade could be improved by $2.5 billion cumulatively from 2004 through 2010, while reducing the need for federal emergency farm assistance payments.

The operation of new ethanol plants could boost employment by 10,000, and the increase in purchases of farm equipment, combined with multiplier effects, could boost employment by another 10,000. Farm income will rise as a result of stronger demand for crops used as feedstocks which, combined with multiplier effects, could boost employment due to off-farm expenditures by an additional 25,000. Significant additional growth will be made possible by technologies under development. The use of residues from corn processing such as corn fiber, and from field residues, could supply additional low-cost sugars for ethanol production, increasing the flexibility of these agricultural biorefineries.

There are also policy-driven opportunities for increased bioenergy use. The excise tax exemption is currently scheduled to decline from the equivalent of 54 cents per gallon of ethanol to 52 cents, and expire September 30, 2007. Policy makers and some members of Congress have made proposals to set, at the federal level, a specific fraction of the electricity to be provided by renewables, of which biopower—second only to hydropower—is a major supplier. Many states have already enacted such renewable portfolio standards for electricity. Similarly, proposals exist for a renewable fuel standard. Federal legislation still needs to be enacted in both cases.

Another way to increase development of renewable technologies that offer environmental benefits is through systems benefit charges. California, Montana, Illinois, Massachusetts, and Rhode Island are implementing this option as electric industry restructuring is impacting development of renewables.

Finally, markets for green electricity are already established and functioning in several jurisdictions as a result of consumer preference for environmentally preferred electricity.

The initiative goals and strategies that follow indicate how we can develop the technologies and remove the barriers to these tremendous opportunities and meet the goal of tripling America's use of biobased products and bioenergy by 2010.

Benefits to the Farmers

From a $2 bushel of corn we can make a combination of fuels, industrial products, chemicals, and feeds with a total wholesale value of more than $8. To benefit from this "added value," farmers increasingly are becoming owners of manufacturing facilities. More than 150 farmer- and cooperatively-owned processing and manufacturing facilities started up in the last 10 years. In Minnesota, 11 of the 14 ethanol refineries, which collectively produce over 90 percent of Minnesota's ethanol needs, are owned directly by more than 8500 Minnesota farmers.

Initiative Goals

There is a strong relationship between markets, technologies, and policies (see Figure 9). At the apex of this relationship are markets for biobased products and bioenergy. This strategic plan describes technology and policy goals that will provide a solid foundation for achieving a national goal of tripling emerging biobased products and bioenergy use by 2010.

But the goals and strategies alone cannot enable us to reach the national goal. Essential for success are the policies to promote R&D and demonstration, and to encourage deployment and commercialization of biobased products and bioenergy. Also essential are the Executive Order 13134 and related orders, industry vision, technology roadmaps and strategic planning processes, and the individual programs of government agencies (see Appendix 1) at both the federal and state levels.

Figure 9. Examples of actions to triple biomass use by 2010. These include research, development, and demonstration in technology and product development. Among actions are coordinated policy modifications to create government markets and facilitate commercial market development. Also included are the continuing implementation of related Executive Orders (13031, Federal Alternative Fueled Vehicle Leadership; 13101, Greening the Government through Waste Prevention, Recycling, and Federal Acquisition; and 13123, Greening the Government through Efficient Energy Management).

Technology Development Goals

GOAL 1

Reduce by two- to ten-fold costs of technologies for integrated supply, conversion, manufacturing and application systems for biobased products and bioenergy by 2010. Technologies include biomass production, collection, transportation, and delivery, as well as biomass conversion technologies to develop fuels, power, heat, chemicals, and materials.

STRATEGIES

- Implement and strengthen coordinated federal R&D programs and justify integrated funding requests across federal departments and agencies.

- Conduct competitive solicitations with the private sector, universities, and others, for which the agencies and government laboratories provide structure, oversight, and technical support on key R&D issues, and ensure that solicitations reach a "critical R&D mass" to achieve the goal.

- Develop resource-efficient, environmentally sound, high productivity feedstock production systems.

Examples of Milestones

- *By 2010, halve the year 2000 cost of producing sugars from lignocellulosic biomass.*

- *By 2010, develop the technologies for cost-competitive biomass gasification platforms for both power and biorefinery co-products.*

- *By 2010, develop 250 new biobased products for commercialization. This number includes at least 20 high-energy use impact biobased products.*

GOAL 2

Demonstrate critical integrated biobased products and bioenergy systems for fuels, heat, power, chemicals and materials between 2002-2008 so they may contribute to the tripling goal by 2010.

STRATEGIES

- Identify the scientific and technological resources that will be needed and fully utilize relevant existing facilities (federal, state, academia, and private).

- Craft public-private sector partnerships to demonstrate technologies.

- Leverage private investments to demonstrate promising technologies.

Examples of Milestones

- *By 2002, complete an inventory of public resources and facilities.*

- *By 2002, demonstrate an integrated, commercial-scale facility for multiple products. One example is to use lignocellulosic biomass to produce sugars, ethanol, and power.*

- *By 2008, demonstrate integrated gasification technologies for producing power and multiple products, including hydrogen. Implement gasification technology in three plants of the pulp and paper industry by 2010.*

- *Between 2002-2008, demonstrate rural-based processing plants for biobased products, including plants owned by farmers or farm cooperatives. Examples include composites, building materials, plastics from crop residues, and small-scale production of power using integrated gasification technology with advanced power sources.*

GOAL 3

Monitor and evaluate the environmental and ecosystem impacts of biobased products and bioenergy systems at all stages of development and apply this information toward improving these systems' safety and environmental benefits.

STRATEGIES

- Identify and foster R&D on biobased products and bioenergy areas that have substantial potential to replace fossil-based fuels, power, heat, chemicals, and materials (including inorganic products replacement) with substantial potential to provide environmental benefits.

- Establish specific review committees with broad public representation and open processes to oversee environmental monitoring and evaluation, in-field biomass production, and facility conversion processes.

- Conduct ongoing life-cycle analyses to evaluate integrated systems and determine areas for environmental improvement.

- Utilize advanced information technologies to collaboratively assemble, analyze, and publicly disseminate information on relevant environmental and ecosystem impacts.

Examples of Milestones

- *By 2002, review environmental and ecosystem monitoring by the federal, state, and local governments agricultural, forestry, and environmental agencies and private sector and non-governmental organizations.*

- *By 2002, develop tools and information resources that will facilitate identification of those biobased products and bioenergy technologies that can provide econom-*

ic, agricultural, energy, and environmental benefits simultaneously and produce a plan to accelerate their development. This will help farmers identify more profitable crops and assist them in developing farm cooperatives to supply biobased products and bioenergy to new markets.

- On an ongoing basis, identify and implement opportunities to leverage these monitoring efforts and expertise, supplementing them as needed, to support life-cycle environmental and ecosystem monitoring of biobased product and bioenergy systems.

- By 2002, establish a national program initiative to promote understanding of the role of bioproducts and bioenergy in enhancing environmental sustainability in USDA's Cooperative Extension Service, in state cooperative extension service programs, and in USDA's NRCS.

- By 2003, evaluate current and prospective life-cycle environmental costs and benefits of key biobased products and bioenergy important to achieve the tripling goal and compare them with fossil-fuel based alternatives.

GOAL 4

Foster innovation-driven science of biomass feedstocks, biobased products, and bioenergy and quickly incorporate these scientific results in the relevant technology development activities.

STRATEGIES

- Strengthen and integrate basic scientific research programs and complementary competitive grant programs across federal agencies and their laboratories, academic institutions, and private-sector firms.

- Enhance human resource development to support scientific R&D programs.

- Strengthen partnerships between the public and private sectors.

- Evaluate biannually the federal, state, and private sector biobased products and bioenergy R&D portfolio to identify gaps in frontier science and technology.

- Identify opportunities for technology transfer from other functional genomics and metabolic engineering R&D, such as on human systems.

- Identify R&D issues that would greatly benefit from dedicated Center of Excellence attention and, where appropriate, extend existing or develop new programs that address key challenge areas.

- Reserve a portion of the R&D funding for high-risk frontier science opportunities to nurture innovation.

- Support research fellowship programs at universities and national laboratories in key science areas that benefit biomass feedstocks, biobased products, and bioenergy.

Examples of Milestones

- By 2010, develop complete functional genomics for five target feedstocks. Use these capabilities, for instance, to double overall lignocellulosic productivity at significantly reduced inputs.

- By 2005, develop complete functional genomics for 10 target biocatalysts and develop associated metabolic engineering by 2010 to triple reaction rates and significantly reduce end-product inhibition.

- On an ongoing basis, create advanced tools for information and computing technologies including those that encourage collaborative learning and intelligence. Use them to develop functional genomics and metabolic engineering science and technology, and to provide specific information on targeted plants and organisms.

Market and Public Policy Goals

GOAL 5

Coordinate policies to achieve early market adoption of biobased products and bioenergy and create demand for biobased products and bioenergy.

STRATEGIES

- Identify in detail the principal barriers to the research, development, demonstration, and deployment of biobased products and of bioenergy and systematically develop coordinated policy mechanisms to overcome them. This may include tax incentives, environmental offsets, risk mitigation mechanisms in early deployment, buy-down mechanisms, and others.

- Identify existing federal and state authorities that can be used to facilitate early adoption of biobased technologies and products.

- Link environmental benefits of biobased products and bioenergy to public policy development.

- Resolve infrastructure, performance, environmental, and health testing issues that present a barrier to the market adoption of biobased products and bioenergy.

- Encourage early development and adoption of standards and labels for biobased products and bioenergy. Work with the private sector and non-governmental organizations to identify the appropriate role of government in this effort.

Examples of Milestones

- *By 2002, develop a coordinated proposal for modifying the tax code and other policies to encourage private-sector investments in integrated biobased products and bioenergy development and deployment and to overcome the barriers limiting development of this sector.*

- *By 2002, develop mechanisms to assist the buy-down of initial capital costs of new technologies, and mitigate risks in deploying first-of-a-kind technologies.*

- *By 2002, develop mechanisms to assist farmers and farmer cooperatives to gain economic benefits from biobased products and bioenergy.*

- *By 2003, announce labeling programs to facilitate marketing of products and technologies. These programs will be defined with the private sector.*

- *By 2005, have federal incentives such as tax credits and appropriate tax-credit programs in place and funded. Have appropriate parallel incentive mechanisms for public power companies.*

- *By 2005, complete performance testing on those biobased products and bioenergy of primary importance to achieving the tripling goal.*

GOAL 6

Increase federal government purchases or production of bioenergy to 5 percent and relevant biobased products purchases to 10 percent by 2010.

STRATEGIES

- Inform consumers and government employees about the benefits of biobased products and bioenergy so they will support the effort.

- Leverage and coordinate with existing "green" purchasing programs, such as EPA's Comprehensive Procurement Guideline program and EPA's Environmentally Preferable Purchasing Program, to ensure that consistent messages about preferential buying requirements are communicated to federal purchasers.

- Facilitate enactment of federal and state legislation to assist purchases of biobased products and bioenergy.

- Use targeted demonstration programs to collect data over time and quantify benefits and costs of biobased products and bioenergy use in federal facilities.

Examples of Milestones

- *By 2002, publish an USDA-approved biobased products list for federal procurement.*

- *By 2002, develop legislative language and facilitate enactment of legislation to modify Section 6002 of the Resource Conservation and Recovery Act to require federal procurement officials to purchase certain levels of biobased products and bioenergy.*

- *In 2010, bioenergy accounts for 5 percent of federal energy facilities and biobased products penetrate 10 percent of federal purchases of relevant products.*

- *By 2010, work with all states to enact legislation that increases state purchases of biobased products and bioenergy.*

GOAL 7

Facilitate tripling the use of emerging biobased products and bioenergy by 2010 in a manner consistent with federal resource conservation and environmental policies.

STRATEGIES

- Collaborate with the appropriate federal, state, and local agencies to facilitate private-sector investment in key areas of biobased products and bioenergy for widespread implementation of technologies. For instance, DOE and EPA together can accelerate the commercial implementation of cofiring and demonstration activities such as gasification of biomass and black liquor.

- Develop science-based education and outreach programs and materials, directed toward classroom teaching and consumer education to explain the environmental sustainability and product performance of biobased products and bioenergy.

Examples of Milestones

- *By 2002, work with educators to develop K-12 classroom materials that focus on the science-based environmental sustainability attributes of biobased products and bioenergy.*

- *By 2002, establish 25 centers of excellence at colleges and universities to develop college-level classroom materials and outreach education materials focused on the role of biobased products and bioenergy in enhancing environmental sustainability.*

- *By 2005, implement cofiring in 5 percent of pulverized coal boilers. Joint DOE and EPA action with public and private electricity generators will achieve this milestone.*

- *By 2010, triple biofuels production through cooperative USDA, DOE, and EPA actions with industry.*

Strategies Common to All Goals:

- Use public-private partnerships to leverage taxpayer dollars to the maximum extent feasible while ensuring that the results of the work are quickly and effectively moved toward commercial products of value to the public. Efforts will be made to leverage international R&D when appropriate and to encourage international export of U.S. biobased products and bioenergy.

- Utilize strategic planning processes with stakeholders such as DOE-facilitated Vision and Technology Roadmaps to help guide the selection of areas for federal investment and bring in diverse stakeholders to jointly develop biobased products and bioenergy.

- Ensure relevance of targeted federal investments—the Biomass Research and Development Technical Advisory Committee, established by section 306 of P.L.106-224, will coordinate programs within and among departments and agencies to maximize the benefits derived from federal grants and assistance and by bringing coherence to federal strategic planning.

- Use external peer review to oversee initiative activities and provide independent validation or correction of the work plans and results.

- Establish appropriate criteria for terminating unsuccessful projects; develop sunset criteria for the phaseout of conventional and biobased product and bioenergy incentives, taking into account direct and indirect supports of competing technologies.

- Employ competition wherever possible and appropriate, such as through competitive solicitations, to ensure effective execution of work. This includes using market-driven approaches for technology-cost buy-down and other public-private activities.

- Pursue multiple benefits wherever possible. These include simultaneously increasing farm and forest income, rural employment, and environmental quality, while reducing oil imports. It also includes identifying ways to capture multiple benefits from public investments.

- Take advantage of technology and information transfer mechanisms set up by the agencies and increase their coordination. The coordination of the member and participating agencies by the Biomass Research and Development Board will accelerate the dissemination of information across agencies and to stakeholders.

- Develop education initiatives to broadly inform consumers and the general public of the science-based environmental sustainability and product-performance benefits associated with the increased use of biobased products and bioenergy systems.

- Establish a mechanism for measuring progress via appropriate metrics tailored for each area, power, fuels, and products. Update measures periodically (for instance, through the Energy Information Administration, other USDA or DOC tracking of specific biobased products and bioenergy industry activities).

Requirements:

To accomplish the goal of tripling the U.S. use of emerging bioproducts and bioenergy, as identified by the President and outlined in this strategic plan, would require R&D and demonstration funding to be increased to $500 to $1000 million per year of public-sector investment. This is roughly 2 to 4 times the current levels of investment. This funding would accelerate technology development. Public funding levels would decrease as technologies penetrate private markets. Together with the policy changes suggested in this plan, this public-sector investment would leverage roughly hundreds of millions of dollars of private-sector investments over the next decade to reach the national goal. The details of these public sector investments are being separately developed in the document entitled Goals, Metrics, Milestones and Resources.

These estimates are similar to those proposed by stakeholders during meetings in the Vision and Technology Roadmap efforts facilitated by DOE. Industry and academia stakeholders suggested that budgets needed to double until they reached $1 billion per year, and then need to be sustained to allow development of the technologies. Such a sustained investment is necessary to achieve the aggressive national goals and to reap the significant benefits of emerging biobased products and bioenergy for the United States' economy, its environment, and its energy security and diversity.

Examples of Actions to Reach the National Goal

To assess and enable the country to accrue all the benefits from the increased use of biobased products and bioenergy, combined actions by several agencies are necessary. Table 2 shows the agencies involved in these actions.

TABLE 2 Agencies Participating, in the Implementation of the National Goals

Goals and Actions	DOE	USDA	EPA	NSF	DOC/NIST	DOI	TREASURY	OSTP	FEE	GSA	TVA
Technology Development Goals											
1. Technology Cost Reduction	X	X	X		X			X			X
2. Critical Technology Demonstration	X	X			X			X	X		
3. Environmental/Ecosystem Impact	X	X	X		X	X		X	X		X
4. Innovative Science & Technology	X	X	X	X	X			X			X
Public Policy and Markets Goals											
5. Coordinated Policies	X	X	X				X	X	X		X
6. Increase Federal Market Pull	X	X	X		X		X	X	X	X	X
7. Increase Commercial Market Pull	X	X	X		X		X	X	X		X
Examples of Actions											
Tax Code Modifications	X	X	X				X	X	X		
New Legislation	X	X	X				X	X			
Financing Barrier Reduction	X	X	X		X		X	X	X		X
Federal Markets	X	X	X			X		X	X	X	
National/International Markets	X	X	X		X			X	X		
Outreach and Communications	X	X	X			X		X	X		X
Codes/Standards	X	X	X		X						
Incentive Payments	X	X						X	X		X
Related Executive Orders: 13101		X	X						X		
Related Executive Orders: 13031, 13123	X		X						X		

Acronyms

DOE	U.S. Department of Energy
USDA	U.S. Department of Agriculture
EPA	Environmental Protection Agency
NSF	National Science Foundation
DOC	Department of Commerce
NIST	National Institute of Standards and Technologies
DOI	Department of Interior
TREASURY	Department of the Treasury
OSTP	Office of Science and Technology Policy
FEE	Federal Environmental Executive
GSA	General Services Administration
TVA	Tennessee Valley Authority

Executive Orders

13031	Federal Alternative Fueled Vehicle Leadership
13101	Greening the Government through Waste Prevention, Recycling and Federal Acquisition
13123	Greening the Government through Energy Efficiency Management

Appendix I. Federal Government Programs Relevant to Biobased Products and Bioenergy and their Web Sites

Departments, Agencies, Offices, Programs	Web Site
U.S. Department of Energy	**http://www.doe.gov/**
Energy Supply Research and Development	
Renewable Energy	
Biomass Power Systems	http://www.eren.doe.gov/biopower/flash.html
Biofuels Energy Systems	http://www.ott.doe.gov/biofuels/
Hydrogen Energy Research and Development	http://www.eren.doe.gov/power/hydrogen.html
Science	***http://www.science.doe.gov***
Basic Energy Sciences	http://www.sc.doe.gov/production/bes/bes.html
Energy Biosciences	*http://www.sc.doe.gov/production/bes/Division.htm#biosciences*
Chemical Sciences	*http://www.sc.doe.gov/production/bes/Division.htm#chemical*
Engineering and Geosciences	*http://www.sc.doe.gov/production/bes/Division.htm#EngGeo*
Biological and Environmental Research	*http://www.sc.doe.gov/production/ober/ober_top.html*
Energy Conservation	***http://www.eren.doe.gov/***
Industry Sector	http://www.eren.doe.gov/EE/industrial.html
Industries of the Future – specific	
Forest Products	*http://www.oit.doe.gov/forest/*
Agriculture	*http://www.oit.doe.gov/agriculture/*
Industries of the Future – crosscutting	
Transportation	http://www.eren.doe.gov/EE/transportation.html
Fuels Utilization R&D	*http://www.ott.doe.gov/otu/afutil.html*
Fossil Energy Research and Development	***http://www.fe.doe.gov/***
Coal and Power Systems	http://www.fe.doe.gov/programs_coalpwr.html
Central Systems	http://www.fe.doe.gov/coal_power/central_power.html
Gasification Cycles	*http://www.fe.doe.gov/coal_power/gasification/index.html*
Turbines	*http://www.fe.doe.gov/coal_power/turbines/index_industrial.html*
Distributed Systems	http://www.fe.doe.gov/coal_power/distributed_power.html
Fuel Cells	*http://www.fe.doe.gov/coal_power/fuelcells/index.html*
Natural Gas	*http://www.netl.doe.gov/scng/*

Departments, Agencies, Offices, Programs | Web Site

Departments, Agencies, Offices, Programs	Web Site
U.S. Department of Agriculture	http://usda.gov
Agriculture Research Service	*http://www.ars.usda.gov/*
Research in Integration of Agricultural Systems and Research Support	http://www.nps.ars.usda.gov/
Natural Resources and Sustainable Agricultural Systems	http://www.nps.ars.usda.gov/
Rangeland, Pasture, and Forages	*http://www.nps.ars.usda.gov/programs/programs.htm?NPNUMBER=205*
Manure and Byproduct Utilization	*http://www.nps.ars.usda.gov/programs/programs.htm?NPNUMBER=206*
Integrated Agricultural Systems	*http://www.nps.ars.usda.gov/programs/programs.htm?NPNUMBER=207*
Crop Production, Product Value, and Safety	http://www.nps.ars.usda.gov/
Plant, Microbial, and Insect Genetic Resources, Genomics, and Genetic Improvement	*http://www.nps.ars.usda.gov/programs/programs.htm?NPNUMBER=301*
Plant Biological and Molecular Processes	*http://www.nps.ars.usda.gov/programs/programs.htm?NPNUMBER=302*
New Uses, Quality, and Marketability of Plant and Animal Products	*http://www.nps.ars.usda.gov/programs/programs.htm?NPNUMBER=306*
Bioenergy and Energy Alternatives	*http://www.nps.ars.usda.gov/programs/programs.htm?NPNUMBER=307*
Federal Technology Transfer to the Private Sector	http://www.ott.ars.usda.gov/
Cooperative State Research, Education and Extension Service	*http://www.reeusda.gov/*
Agricultural Materials	http://www.reeusda.gov/1700/programs/agri.htm
National Research Initiative	http://www.reeusda.gov/ree/nri/
Small Business Innovative Research	http://www.reeusda.gov/sbir/
McIntire-Stennis Cooperative Forestry	http://www.reeusda.gov/1700/legis/mcstenni.htm
Payments under the Hatch Act	http://www.reeusda.gov/crgam/oep/fmbstaff.htm http://www.reeusda.gov/1700/legis/hatch.htm
Payments to 1890 Colleges and Tuskegee	http://www.reeusda.gov/1700/programs/1890.htm
Special Research Grants	http://www.reeusda.gov/ree/reedir/dgz00662.htm
Initiative for Future Agriculture and Food Systems	http://www.reeusda.gov/1700/programs/IFAFS/IFAFS.htm
Forest Service	*http://www.fs.fed.us/research*
Forest and Rangeland Research	http://www.fs.fed.us/research/rvur/
Vegetation Management and Protection Research	http://www.fs.fed.us/research/vmpr
Forest Products	http://www.fs.fed.us/research/rvur/products/index.htm
Natural Resources Conservation Service	*http://www.nrcs.usda.gov/*
Financial and Technical Assistance for Conservation	http://www.nrcs.usda.gov/NRCSProg.html
Economic Research Service	*http://www.ers.usda.gov/*
Animal and Plant Health Inspection Service	*http://www.aphis.usda.gov*
Rural Development – grants and loans	*http://www.rurdev.usda.gov/*
Office of Energy Policy and New Uses	*http://www.usda.gov/agency/oce/oepnu/index.htm*
Commodity Credit Corporation – market development	*http://www.fsa.usda.gov/daco/bioenergy/bioenergy.htm*

Departments, Agencies, Offices, Programs	Web Site
National Science Foundation	http://www.nsf.gov/
Biological Sciences	*http://www.nsf.gov/home/bio/*
Biological Infrastructure	http://www.nsf.gov/bio/dbi/
Environmental Biology	http://www.nsf.gov/bio/deb/
Integrative Biology and Neuroscience	http://www.nsf.gov/bio/ibn/
Molecular and Cellular Biosciences	http://www.nsf.gov/bio/mcb/
Engineering	*http://www.nsf.gov/home/eng/*
Bioengineering and Environmental Systems	http://www.eng.nsf.gov/bes/
Chemical and Transport Systems	http://www.eng.nsf.gov/cts/
Design, Manufacture, and Industrial Innovation	http://www.eng.nsf.gov/dmii/
Mathematical and Physical Sciences	*http://www.nsf.gov/home/mps/*
Chemistry	http://www.nsf.gov/mps/chem/
Materials Research	http://www.nsf.gov/mps/dmr/
Mathematical Sciences	http://www.nsf.gov/mps/dms/
Environmental Protection Agency	http://www.epa.gov
Industry Partnerships, Project XL	*http://www.epa.gov/ProjectXL/*
Methane Energy	*http://www.epa.gov/methane/*
Landfill Methane Outreach	http://www.epa.gov/lmop/
AgStar Partnership (joint with USDA and DOE)	http://www.epa.gov/outreach/agstar/
Office of Research and Development	*http://www.epa.gov/ORD/*
Environmental Technology Verification	http://www.epa.gov/etv/
Office of Pollution Prevention and Toxics	*http://www.epa.gov/internet/oppts/*
Green Chemistry	http://www.epa.gov/opptintr/greenchemistry/program.htm
Genetically Modified Microorganisms	http:/www.epa.gov/opptintr/biotech
Plant Pesticides	http://www.epa.gov/pesticides/biopesticides
Extramural Research and Development	*http://www.epa.gov/AthensR/extrmural/index.html*
Comprehensive Procurement Guidelines	*http://www.epa.gov/cpg*
Department of Commerce	http://www.doc.gov
Advanced Technology	http://www.atp.nist.gov/
Tennessee Valley Authority	http://www.tva.gov/
Renewables and Biomass	
Public Power Institute	http://www.publicpowerinstitute.org

Appendix II. Agencies and their Missions

Mission	Agency
Supporting production agriculture; ensuring a safe, affordable, nutritious, and accessible food supply; caring for agricultural, forest, and rangelands; supporting sound development of rural communities; providing economic opportunities for farm and rural residents; expanding global markets for agricultural and forest products and services.	**Department of Agriculture** *"A healthy and productive Nation in harmony with the land"*
Working to assure clean, affordable, and dependable supplies of energy for our nation, now and in the future.	**The Department of Energy** *Science, Security and Energy: Powering the 21st Century*
Realizing the promise of the 21st century depends in large measure on today's investments in science, engineering and mathematics research and education. NSF investments are in people, their ideas and the tools they use. NSF catalyzes the strong progress in science and engineering needed to secure the Nation's future.	**The National Science Foundation** *Enabling the Nation's future through discovery, learning and innovation*
Protecting public health and safeguarding the natural environment—air, water, and land—upon which life depends.	**United States Environmental Protection Agency** *"...to protect human health and to safeguard the natural environment..."*
Cooperating with partners to extend the benefits of natural and cultural resource conservation and outdoor recreation throughout this country and the world.	**Department of the Interior** *To protect and provide access to our Nation's natural and cultural heritage and honor our trust responsibilities to tribes*

Agency	Mission
Office of Science and Technology Policy	Serve as a source of scientific and technological analysis and judgment for the President with respect to major policies, plans, and programs of the Federal Government.
Department of Commerce *National Institute of Standards & Technology*	Serving the needs of all Americans, creating job opportunities for American workers, and enhancing the competitiveness of United States industry in the global marketplace.
Environmental Executive *Task Force on Greening the Government through Waste Prevention and Recycling*	Expanding and strengthening the Federal government's commitment to recycling and buying recycled content and environmentally preferable products.
The Office of Management and Budget	Building on the success of fiscal discipline, improving performance through better management, strengthening the nation in the 21st century, and building prosperity for the future.
The Department of the Treasury	Maximizing the potential of the U.S. and world economies for growth and stability.
Tennessee Valley Authority *The Power of the Public Good*	TVA is a unique federal corporation that is the nation's largest public producer of electricity, the steward of the Tennessee River system, and a regional economic development agency.

Appendix III. Glossary of Terms

absorbent: a material that, because of surface retention, accumulates layers of molecules or atoms of another material (usually a gas or liquid) with which it is in contact. For example, because of its large surface area, charcoal can serve as an absorbent for large volumes of gas.

acetic acid: (CH_3COOH) - an inexpensive organic acid that is a component of vinegar and that is an important starting substance for making textile fibers, vinyl plastics, polyesters, and other chemicals.

acids: a class of chemical compounds capable of transferring hydrogen ions in solution, that turn blue litmus paper to red, that react with certain metals to form salts, and that react with bases to form salts. Acids have a wide range of industrial uses and are essential to life itself.

alcohols: any of a class of chemical compounds (such as methanol, ethanol, propanol) that contains the hydroxyl group, OH. Alcohols can be used for a variety of applications that range from fuel, to beverage, to industrial solvent, to a chemical intermediate used for manufacturing a wide variety of chemicals and materials.

amylase: an enzyme that hydrolyzes (decomposes with water) reserve carbohydrates, starch in plants, and glycogen in animals.

aromatic compounds: a large group of organic chemicals that usually contain closed rings of carbon atoms. About half of all known organic compounds are aromatic.

biobased products: commercial or industrial products, other than food and feed, derived from biomass feedstocks. Biobased products include green chemicals, renewable plastics, natural fibers, and natural structural materials. Many of these products possess unique properties unmatched by petroleum-based products or can replace products and materials traditionally derived from petrochemicals. However, new and improved processing technologies will be required.

biobased technologies: those technologies that use biomass feedstocks as the raw material for making products or for producing energy.

biocatalyst: usually refers to enzymes and microbes, but it can include other catalysts that are living or that were extracted from living organisms, such as plant or animal tissue cultures, algae, fungi, or other whole organisms.

biodegradable: the attribute of a substance that can be broken down into simpler compounds by microorganisms.

biodiesel: a biofuel produced through a process in which organically derived oils are combined with alcohol (ethanol or methanol) in the presence of a catalyst to form ethyl or methyl ester. The biomass-derived ethyl or methyl esters can be blended with conventional diesel fuel or used as a neat fuel (100% biodiesel). Biodiesel can be made from soybean or rapeseed oils, animal fats, waste vegetable oils, or microalgae oils.

bioeconomic: an economy based on biological sciences and advances in related engineering disciplines and physical sciences.

bioenergy: the energy contained in material produced by photosynthesis (including organic waste) may be used directly or indirectly to manufacture fuels and substitutes for petrochemicals and other energy-intensive products. The production of energy from biomass, for example, can be direct (e.g., via combustion) or indirect (e.g., via conversion into ethanol or through gasification).

biofuels: fuels made from biomass resources, including the liquid fuels ethanol, methanol, biodiesel, Fischer-Tropsch diesel, and gaseous fuels such as hydrogen and methane. Conversion of biomass to fuels generally involves conversion to an intermediate (sugar or syngas) and then to a fuel by a catalyst.

biomass: organic matter available on a renewable basis. Biomass includes forest and mill residues, agricultural crops and residues, wood and wood residues, animal wastes, livestock operation residues, aquatic plants, fast-growing trees and plants, and the organic portion of municipal and relevant industrial wastes.

biomass gasification: biomass is heated without excess of air so that instead of completely being combusted, an intermediate gas is formed. This gas has hydrogen and carbon monoxide, in addition to carbon dioxide and water, and small amounts of other products. The resulting gas can then be burned to produce power or processed into chemicals and fuels.

biomass hydrolysis: the use of an aqueous solution (often acidic) to decompose biomass into simpler chemical units. Alternatively, enzymes can be used. Complete hydrolysis breaks biomass into its simplest units (monomers), which can serve as repeated building blocks for making polymers, alcohols, acids, fuels, plastics, and other products.

biomass residues: all forms of biomass yield byproducts and waste streams that have significant potential for energy or further processing. For example, making solid wood products and pulp from logs produces bark, shavings and sawdust, and spent pulping liquors. Since these residues are already collected at the point of processing, they can be convenient and relatively inexpensive sources of biomass for energy or processing into other products. Construction and demolition wood is another example of residues used to make electricity.

biomass resource: any plant-derived organic matter that is available for food, feed, fiber, energy, or biobased products on a sustainable basis. Includes agricultural food and feed crops, agricultural crop wastes and residues, wood wastes and residues, herbaceous and woody energy crops, aquatic plants, animal wastes, some municipal wastes, and other waste materials.

biopower: the use of biomass feedstock to produce electric power, through direct combustion of the feedstock, through gasification and then combustion of the resultant gas, or through other thermal conversion processes. Power is generated with engines, turbines, fuel cells, or other equipment.

biosynthesis gas: a gas, created by gasification of organic material, comprised primarily of hydrogen and carbon monoxide. This gas can be combusted for energy or used to make alcohols, acids, fuels, plastics, and other products.

carbohydrate: any group of organic compounds composed of carbon, hydrogen, and oxygen, including sugars, starches, and celluloses.

carbon - C: the sixth lightest element of the periodic table. Carbon is the basic element that defines organic material and is essential to all life. Carbon compounds constitute greater than 90 percent of the millions of compounds known to exist.

carbon-based industries: industries whose products are derived primarily from biomass or fossil feedstock. This comprises a wide range of industries, including forestry and agriculture, which are sources of renewable carbon and; petroleum, coal, and natural gas, which are sources of fossil carbon.

carbon dioxide: a colorless, odorless noncombustible gas with the formula CO_2 that is present in the atmosphere. It is formed by the combustion of carbon and carbon compounds or by respiration, which is a slow combustion in animals and plants. It is also generated by the gradual oxidation of organic matter in the soil.

carbon sequestering: the physical, chemical, or biological storage of carbon so that it does not reach the atmosphere for centuries or longer. Methods of sequestering include physical storage of carbon dioxide in depleted oil or gas reservoirs and biological storage by forests and biomass plantations.

catalyst: a chemical substance that increases the rate of chemical reaction without being consumed. Biological substances carrying catalytic functions are known as biocatalysts.

catalytic cracking: the use of catalysts and heat to break hydrocarbons into shorter chains of hydrocarbons. Used in the petroleum industry to derive a variety of useful products from crude oil.

cellulase: any of a group of extracellular enzymes that hydrolyze (decompose with water) cellulose.

cellulose: The main carbohydrate in living plants. Cellulose forms the skeletal structure of the plant cell wall.

cofiring: the use of two or more different fuels (e.g. wood residue and coal) simultaneously in the same combustion chamber of a power plant.

depolymerization: the decomposition of macromolecular compounds into relatively simple compounds.

enzymatic hydrolysis: a process by which enzymes (biological catalysts) are used to break down polymers into their components.

enzyme: a protein functioning as a biological catalyst. Enzymes accelerate (often by several orders of magnitude) chemical reactions that would proceed imperceptibly or not at all in their absence.

ethanol - C$_2$H$_5$OH: a colorless liquid that is the product of fermentation used in alcoholic beverages, industrial processes, and as a fuel additive. Also known as grain alcohol.

ethanol-gasoline blend: a blend of unleaded gasoline with ethanol (usually in a ratio of 9 to 1). Used as a motor vehicle fuel to reduce emissions of carbon monoxide, especially in areas that are out of compliance with requirements of the Clean Air Act and its amendments.

ethylene - C$_2$H$_4$: the simplest of the alkenes, a class of compounds that have one or more carbon-carbon double bonds. Ethylene, which is also a natural product of plant respiration and contributes to ripening of fruit, is one of the most important raw materials used in the organic chemical industry. It is used for making polyethylene, antifreeze, ethyl alcohol (ethanol), and other chemicals.

fermentation: the decomposition of organic material to alcohol, methane, etc., by organisms, such as yeast or bacteria, usually in the absence of oxygen.

Fischer-Tropsch process: a process whereby synthesis gas (primarily hydrogen and carbon monoxide) are reacted in the presence of an iron or cobalt catalyst to produce such products as methane, synthetic gasoline, waxes, and alcohols. When used with biomass feedstock, synthetic gasoline derived with the Fischer-Tropsch process is sulfur free.

fossil fuels: solid, liquid, or gaseous fuels formed in the ground after millions of years by chemical and physical changes in plant and animal residues under high temperature and pressure. Oil, natural gas, and coal are fossil fuels.

fuel cell: an electrochemical device that converts the chemical energy of a fuel directly to electricity and heat, without combustion.

genomics: the study of genes and their functions.

glucose - C$_6$H$_{12}$O$_6$: a monosaccharide sugar found in honey and fruits. It is formed by the hydrolysis of more complex carbohydrates, including sucrose, maltose, cellulose, and starch. It is used as a sweetener, in tanning, in making ethanol, and in medicines.

glycols: a family of colorless alcohols used for making a variety of products, including antifreeze, brake fluids, explosives, and solvents. Some glycols such as glycerin are useful for food and medicinal applications.

greenhouse gases: gases that are transparent to solar radiation but that reflect infrared radiation (long-wave radiation produced by the solar-heated Earth) to trap heat in the Earth's atmosphere. Greenhouse gases include water vapor, carbon dioxide, methane, ozone, chlorofluorocarbons, and nitrous oxide.

hydrolysis: decomposition of a chemical compound by reaction with water.

inorganic: compounds that do not contain carbon as the principal element, excepting carbonates, cyanides, and cyanates. Matter other than plant or animal.

lactic acid - $CH_3CHOHCOOH$: a colorless acid that is prepared by the fermentation of starch, cane sugar, or whey, or of lactose in milk. Used in preparing cheese, sauerkraut, soft drinks, and other food products. Lactic acid is also a starting compound for polymers.

lignin: an amorphous phenolic polymer related to cellulose that together with cellulose forms the cell walls of woody plants and acts as the bonding agent between cells. Trees, for example, which have to stand tall, require strong cell walls and therefore have a high lignin content. Lignin, which is a byproduct of converting the cellulosic portion of biomass to products such as cellulose or ethanol, can go through depolymerization to form aromatic compounds. These aromatic compounds as well as ethanol can serve as valuable octane or octane enhancers in gasoline.

lignocellulosic biomass: plants (such as trees) whose primary constituents are compounds of lignin, cellulose, and hemicelluloses, which comprise the essential part of woody cell walls.

metabolic engineering: an emerging research field that combines complex biochemical networks with biochemistry, enzymology, and molecular biology to identify and alter processes within a cell, especially those that limit the productivity of the basic amino acid lysine.

methanol - CH_3OH: the simplest of all alcohols. It is used for the production of formaldehyde (used in plastics, germicides, and fungicides); for the production of other chemicals; and for jet fuels, antifreeze, solvents, as a gasoline additive, and as a denaturant.

methyl tertiary butyl ether - MTBE: an ether compound used as a blending component to raise the oxygen content of gasoline. MTBE—a volatile, flammable and colorless liquid that dissolves rather easily in water—is manufactured by the chemical reaction of methanol and isobutylene.

monosaccharide: a carbohydrate that cannot be hydrolyzed to a simpler carbohydrate.

natural fibers: biomass fibers that are used to make paper products, some textile products, such as from cotton, and many types of rope, twine, and string. New products are being developed based on natural fibers including insulation, structural materials, reinforcing fibers for plastics or other composite materials, and geotextiles for soil erosion-control applications. These fibers are also being used to replace nonrenewable materials as fillers for many products.

natural gas: a hydrocarbon gas obtained from underground sources, often in association with petroleum and coal deposits. It generally contains a high percentage of methane, varying amounts of ethane, and inert gases. Natural gas is used as a heating fuel, for generating electricity, and for making a wide range of chemicals and fuels through Fischer-Tropsch and other processes.

octane number: a numerical rating (from 0 to 100) of the tendency of a fuel to knock when used in an internal combustion engine under standard conditions. The higher the number, the better the antiknock characteristics of the fuel.

oil refinery: an installation where crude oil is cracked and various fractions of oil (which have different weights, boiling points, and condensation points) are separated by distillation and treated to provide many different petroleum products.

organic: chemical compounds based on carbon chains or rings and also containing hydrogen.

oxygenate: a gasoline fuel additive such as ethanol or MTBE that adds extra oxygen to gasoline to reduce carbon monoxide pollution produced by vehicles.

pharmaceuticals: drugs and medicinal products.

photosynthesis: process in which green plants utilize the energy of sunlight to manufacture plant material from carbon dioxide and water in the presence of chlorophyll.

polylactide: a biodegradable thermoplastic derived from lactic acid and that resembles clear polystyrene. It can be processed into fibers and films, thermoformed or injection molded, or used for compost bags, plant pots, diapers, and packaging.

polyethylene: a widely used plastic. It is a polymer of ethylene that is produced at high pressures and temperatures in the presence of any one of several catalysts, depending on the desired properties for the finished product.

polymer: substance consisting of large molecules that are made of many small, repeating units called monomers. Most of the organic substances found in living matter, such as protein and wood, are made of polymers. Many synthetic materials, such as plastics, are also polymers.

polymerization: the process of forming long chains of molecules from simpler molecules.

polysaccharide: a carbohydrate composed of one or many monosaccharides.

primary energy: all energy consumed by end users, excluding electricity but including the energy consumed to generate electricity. This definition becomes important when describing end-use energy consumption. Primary energy consumption is the sum of all primary energy (e.g., coal, petroleum, natural gas, nuclear, hydro, geothermal, solar, wood, wind, etc.) consumed including primary energy consumed to generate electricity. End-use energy consumption is the sum of all primary energy consumed by the end-use sectors (i.e., residential, commercial, industrial, and transportation) plus the electricity consumed by the end-use sectors but excluding the primary energy consumed to generate electricity.

PURPA : Public Utility Regulatory Policy Act. A 1978 law that was, in part, the start of deregulation of the electric power industry. The law requires electric utilities to purchase electricity produced from qualifying power producers that use renewable energy resources or are cogenerators. Utilities were required to purchase power at a rate equal to the avoided cost of generating the power themselves.

reformulated gasoline: gasoline, the properties and composition of which have been formulated to meet the emission requirements of the U.S. Environmental Protection Agency under Section 211 (k) of the Clean Air Act.

renewable (energy) resource : a resource replenished continuously or that is replaced after use through natural means.

residual crude oil: the heaviest fraction of crude oil remaining after distillation of crude oil to remove low-boiling products, straight-run gasoline, and distillate fuels. It is generally the tower bottoms from vacuum distillation.

roadmap: a plan that details the activities and schedule necessary to manage risks, meet opportunities, and identify capabilities required to complete a complex project.

sorbitol - $C_6H_{14}O_6$: an alcohol that occurs naturally and is also produced by the reduction of glucose in aqueous solution. It is used as a plasticizer in certain adhesives, as a softening agent in textiles, paper, and leather, and as a low-calorie sweetener in a variety of food products.

starch: white, odorless, tasteless carbohydrate powder that plays a vital role in the biochemistry of both plants and animals, which convert starch to glucose for energy. Commercially, starch is made from corn or potatoes. It is used to make corn syrup and corn sugar, to sweeten foods, and in the paper and textile industries.

stakeholder: any person or group that derives benefit from or incurs cost in an organization.

steam reforming: a thermal process that uses steam to convert hydrocarbons (primarily in the form of natural gas or coal) into synthesis gas (a mixture primarily of carbon monoxide and hydrogen), typically in the presence of a catalyst. Byproducts of this process are water vapor, carbon dioxide, and a small amount of residual tars.

sugars: a number of chemical compounds in the carbohydrate group that are readily soluble in water; are colorless, odorless, and usually crystallizable; and are more or less sweet in taste. In general, monosaccharides and disaccharides are termed sugars. Sugars are not only used in foods but also are raw material for producing alcohols, acids, and other products.

surfactant: a surface active compound that, when dissolved in water or aqueous solution, reduces its surface tension or the interfacial tension between it and another liquid. There are more than 1,200 commercial surfactants, used for soaps, detergents, wetting agents, and foaming agents.

Taxol: an anticancer drug used for treating ovarian (and possibly other) cancers. Extracted from the bark of the Pacific yew (a type of evergreen), Taxol inhibits breakdown of cell microtubules, consequently preventing cell division.

terephthalate: derived from terephthalic acid $(C_6H_4COOH)_2$, used to make polyester resins for fibers, films, tire cord, apparel, food packaging, beverage bottles, and a variety of other products.

turbine: a rotary engine that converts the energy of a moving stream of water, steam, or gas into electrical or mechanical energy. Today, turbine-powered generators produce most of the world's electrical energy.

watt: the common base unit of power in the metric system. One watt equals one joule per second, or the power developed in a circuit by a current of one ampere flowing through a potential difference of one volt. One Watt = 3.413 Btu/hr. A kilowatt is equal to 1,000 watts and a megawatt is equal to 1 million watts.

xanthan gum: a water-soluble natural gum produced by the fermentation of glucose with certain microorganisms. Used as a binder, extender, or stabilizer in foods and other products.